A COMPLETE GUIDE
TO THE HISTORY AND
MANUFACTURE OF
GRANDFATHER
CLOCKS

British Library Cataloguing-in-Publication Data
A catalogue record for this book is available from the
British Library

Contents

A History of Clocks and Watches

Horology (from the Latin, Horologium) is the science of measuring time. Clocks, watches, clockwork, sundials, clepsydras, timers, time recorders, marine chronometers and atomic clocks are all examples of instruments used to measure time. In current usage, horology refers mainly to the study of mechanical time-keeping devices, whilst chronometry more broadly included electronic devices that have largely supplanted mechanical clocks for accuracy and precision in time-keeping. Horology itself has an incredibly long history and there are many museums and several specialised libraries devoted to the subject. Perhaps the most famous is the *Royal Greenwich Observatory,* also the source of the Prime Meridian (longitude 0° 0' 0"), and the home of the first marine timekeepers accurate enough to determine longitude.

The word 'clock' is derived from the Celtic words *clagan* and *clocca* meaning 'bell'. A silent instrument missing such a mechanism has traditionally been known as a timepiece, although today the words have become interchangeable. The clock is one of the oldest human interventions, meeting the need to consistently measure intervals of time shorter than the natural units: the day, the lunar month and the year. The current sexagesimal system of time measurement dates to approximately 2000 BC in Sumer. The Ancient Egyptians divided the day into two twelve-hour periods and used large obelisks to track the movement of the sun. They also developed water clocks, which had also been employed frequently by the Ancient Greeks, who called them 'clepsydrae'. The Shang Dynasty is also believed to have used the outflow water clock around the same time.

The first mechanical clocks, employing the verge escapement mechanism (the mechanism that controls the rate of a clock by advancing the gear train at regular intervals or 'ticks') with a foliot or balance wheel timekeeper (a weighted wheel that rotates back and forth, being returned toward its centre position by a spiral), were invented in Europe at around the start of the fourteenth century. They became the standard timekeeping device until the pendulum clock was invented in 1656. This remained the most accurate timekeeper until the 1930s, when quartz oscillators (where the mechanical resonance of a vibrating crystal is used to create an electrical signal with a very precise frequency) were invented, followed by atomic clocks after World War Two. Although initially limited to laboratories, the development of microelectronics in the 1960s made quartz clocks both compact and cheap to produce, and by the 1980s they became the world's dominant timekeeping technology in both clocks and wristwatches.

The concept of the wristwatch goes back to the production of the very earliest watches in the sixteenth century. Elizabeth I of England received a wristwatch from Robert Dudley in 1571, described as an arm watch. From the beginning, they were almost exclusively worn by women, while men used pocket-watches up until the early twentieth century. This was not just a matter of fashion or prejudice; watches of the time were notoriously prone to fouling from exposure to the elements, and could only reliably be kept safe from harm if carried securely in the pocket. Wristwatches were first worn by military men towards the end of the nineteenth century, when the importance of synchronizing manoeuvres during war without potentially revealing the plan to the enemy through signalling was increasingly recognized. It was clear that using pocket watches while in the heat of battle or while mounted on a horse was impractical, so officers began to strap the watches to their wrist.

The company H. Williamson Ltd., based in Coventry, England, was one of the first to capitalize on this opportunity. During the company's 1916 AGM it was noted that '...the public is buying the practical things of life. Nobody can truthfully contend that the watch is a luxury. It is said that one soldier in every four wears a wristlet watch, and the other three mean to get one as soon as they can.' By the end of the War, almost all enlisted men wore a wristwatch, and after they were demobilized, the fashion soon caught on - the British *Horological Journal* wrote in 1917 that '...the wristlet watch was little used by the sterner sex before the war, but now is seen on the wrist of nearly every man in uniform and of many men in civilian attire.' Within a decade, sales of wristwatches had outstripped those of pocket watches.

Now that clocks and watches had become 'common objects' there was a massively increased demand on clockmakers for maintenance and repair. Julien Le Roy, a clockmaker of Versailles, invented a face that could be opened to view the inside clockwork – a development which many subsequent artisans copied. He also invented special repeating mechanisms to improve the precision of clocks and supervised over 3,500 watches. The more complicated the device however, the more often it needed repairing. Today, since almost all clocks are now factory-made, most modern clockmakers *only* repair clocks. They are frequently employed by jewellers, antique shops or places devoted strictly to repairing clocks and watches.

The clockmakers of the present must be able to read blueprints and instructions for numerous types of clocks and time pieces that vary from antique clocks to modern time pieces in order to fix and make clocks or watches. The trade requires fine motor coordination as clockmakers must frequently work on devices with small gears and fine machinery, as well as an appreciation for the original art form. As is evident from this very short history of clocks and watches, over the centuries the items themselves have changed – almost out of recognition, but the importance of time-keeping has not. It is an area which provides a constant source of fascination and scientific discovery, still very much evolving today. We hope the reader enjoys this book.

THE DEVELOPMENT OF GRANDFATHER CLOCK CASES.

Before any design, in furniture, attains to the dignity of a fashion, it is necessary for development to proceed on stated lines, and in definite quantities. No example which is unique, that is, has no resemblance to any other, can be described as being in any fashion. The resemblance between one piece and another may be only general, but there must be similarity enough to establish a style.

It follows, therefore, that the long-case clock, in its earliest stages, conformed to no style and inaugurated no fashion. Within a short space of time, however, certain definite types arose, differing, in details, with each maker, yet having a generic resemblance. What is more, there are points of similarity between the long-case and its smaller brother, the table clock. Perhaps the earliest type is the portico top, an architectural form with the classical moulding sections which are nearly always found on these cases. In nearly every instance, the hoods of the early long-cases were fitted with tongued-and-grooved runners at the back, to slide upwards to obtain access to the hands for winding and adjusting. At the outer corners of these hoods, three-quarter twisted columns were the rule, sometimes "responded" by quarter columns of similar design on the back corners.

A twisted column has, naturally, a slope to the spiral, to right or left, respectively. It would appear natural to balance this on a flanking pair of columns, one to the right and one to the left, but there appears to have been no rule in this respect; I have seen as many where the columns are right and left-handed, as others where the inclination of both is in the same direction. These spiral columns appear to have been a general fashion until about 1705, but, later on, these flanking balusters are plain-shafted, often with bases and caps of brass. The latter had a definite meaning when it became the custom to open the front door instead of sliding the entire hood upwards. The hinging of these doors was by pivots, attached to the brass abacus and base of these columns with plain shafts, as the door had to swing clear to avoid contact between the column and the framing. These pivots or pegs would have broken away if the abacus and base had been made of wood.

FIG. 45.

Enlarged cresting of the clock, fig. 108.

The portico-top cases appear to have been nearly always veneered with ebony or lignum vitae, or stained black on pear-wood. Obviously, there are certain parts of a case where a rare wood, in the solid, such as ebony, could not be employed. The use of these two woods, in itself, suggests a foreign origin for these early cases. It is extremely rare to find a portico-top case in either plain walnut or marqueterie. I have never seen one example. Shortly after 1680, if not just before, these cases appear to have been standardised in type, at least by the important makers. With veneers of plain walnut, or floral marqueterie in panel, the hoods were either square at the top, or with an inverted "cushion-moulded" superstructure. Alternating with this form, rather than replacing it, was a carved cresting of walnut, the design of a scrolled pediment centring in a shell or a cherub's head, and surmounted by a simple knob. Joseph Knibb seems to have used this form in nearly all his better clocks.

The early marqueterie, of floral forms in yellow wood, alternating with white and green-stained ivories, and always in panels surrounded by walnut, usually in "oyster-pieces" (*i.e.*, saplings cut into slices across the grain), is of decidedly Dutch character. Another inlay device was one of stars in alternating light and dark woods, to give a false effect of relief, and this is also foreign in conception. This Dutch character disappears after about 1685, and the later cases are typically English in design, general appearance, and workmanship.

FIG. 46.

Enlarged cresting of the clock, fig. 106.

With all the early cases, up to about 1705, there is one characteristic detail which never varies: the quarter-round section of the moulding just under the hood. At a later date this is changed to a hollow, or cavetto, and this may be taken, positively, as an indication of an early or a later date respectively. It was customary to use the same moulding, in reverse, for the junction of the trunk with the base, but this is not invariable; sometimes this moulding consists of several members.

Another feature which persists with some makers, throughout the whole of the eighteenth century, but is never used by others, is a glazed lenticle or aperture, in the trunk door, through which the pendulum-bob can be seen.

Its use was, obviously, to see, at a glance, if the clock was going, and, in the absence of a second's dial, would have a definite function. From the point of appearance, however, it had one drawback. Standing in front of the clock, this lenticle was much below eye-level, and, in consequence, it had the illusion of being above the pendulum-bob, instead of opposite to it, the pendulum being some six inches, or more, back from the trunk door.

FIG. 47. Enlarged cresting of the clock, fig. 90.

FIG. 48.

Enlarged cresting of the clock, fig. 88.

The era of plain walnut is not unbroken. There are intervals of marqueterie, either in panel or "all-over,"— culminating in intricate patterns of fine scrolling—and lacquer (the latter a very short and highly intermittent fashion), then we get a reversion again to plain walnut in the hands of makers such as the Ellicotts. Mahogany appears to become general, for furniture, shortly after 1730, but not for clockcases. The reasons for this will be given in a later chapter. It is sufficient to state here that a mahogany long-case rarely dates before 1760.

After about 1725, the arch form supersedes the older square dial, and remains in vogue, with the exception of precision or "regulator" clocks, during the whole of the eighteenth century, and after. Case designs vary, extremely, in this arch-dial period. The fashion for marqueterie largely declined at this date, so it is exceptional to find an arch-dial clock in an inlaid case, although examples have been known. In nearly every instance, however, they exhibit signs of later reconstruction, probably done shortly after they were made.

Such a case, however, is very different to one which has been altered with intent to deceive, and, in recent years, to enhance its rarity and value. Clocks in these cases are usually by makers far too late to have been possible in conjunction with marqueterie, which was wholly out of fashion in 1735, if not earlier.

FIG. 49.

Enlarged cresting of the clock, fig. 96.

Very early long-cases are found, sometimes, without plinths to the trunks. Figs. 3, 88, 92, 94, 98, 108 and 132 are examples. For many years I had the idea that these cases had been made to stand in moulded trays with adjustable feet, for the better setting up of the clock. In 1927 I found an early Fromanteel clock in Philadelphia, formerly the property of Benjamin Franklin, with its loose tray intact, as it was made. These trays, not being fixed to the cases, would tend to get lost, in fact, the tray on this Franklin clock is the only original one I have found, so far.

It would be strange, indeed, if there were no relation between long-case and table, or "bracket," clocks, as both were often made in the same workshop. The consideration of these, and their illustration, must be deferred to a later chapter.

————◇————

Note.—The illustrations on the following pages have been kept in the order of makers, tracing the development of each for easy reference. The style progressions and dates, therefore, continually recur, but this is unavoidable. The same plan has been followed with mantel clocks later on.

FIG. 50.

"A. FROMANTEEL, LONDINI, FECIT"

Eight-day striking clock, bolt-and-shutter maintaining. power. Bob pendulum, lift-up hood. Capitals and mounts finely chased brass, mercury gilt, ebony veneered case.

Total height 6 ft. 2 in., 8 1/4 in. dial.

c. 1665

FIG. 51.

Enlarged dial of fig. 50.

FIG. 52.

"JOHANNES FROMANTEEL, LONDINI, FECIT"

Oak case, veneered ebony.

6 ft. 0 1/4 in. total height.

c. 1665

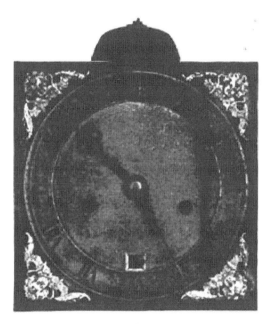

FIG. 53.

FIG. 54.

FIGS. 53 and 54.

Enlarged dial and view of back of movement of Fig. 52.

Eight-day striking clock, silver hour circle to dial. Early
pierced hands.

Bolt-and-shutter maintaining power.
8 1/2 in. dial.

JOHN FROMANTEEL

FIG. 55.

"JOHANNES FROMANTEEL LONDINI. FECIT"

Eight-day striking clock, bob pendulum. Bolt-and-shutter maintaining power. Lift-up hood. Case of oak, veneered ebony. 6 ft. 1 in. total height. 8 5/8 in. dial.

c. 1665

FIG. 56.

Enlarged dial of fig. 55.
EDWARD EAST

FIG. 57.

"EDUARDUS EAST. LONDINI"

Eight-day striking clock. Bob pendulum; outside locking-plate. Bolt-and-shutter maintaining power. Case of oak, veneered walnut and inlaid star marqueterie and stringing.

Lift-up hood. 5 ft. 5 in. high. 7 in. dial.

c. 1665

FIG. 58.
Enlarged dial of fig. 57.

FIG. 59.

"EDUARDUS EAST; LONDINI"

(c.c. 1631. Master 1645 & 1652)

Eight-day striking clock. Bolt-and-shutter maintaining power. Case of oak, veneered walnut and inlaid floral marqueterie in panels. Lift-up hood. 6 ft. 5 1/4 in. high. 10 in. dial.

c. 1680

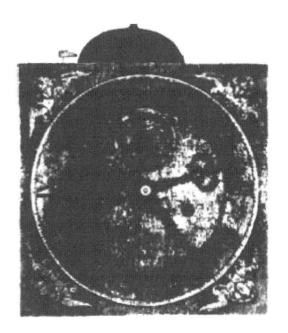

*FIG. 60.*Enlarged dial of fig. 59.

WILLIAM CLEMENT

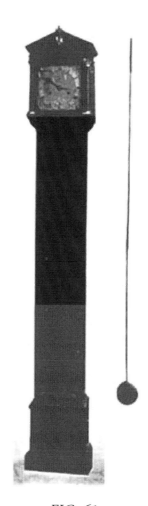

FIG. 61.

"GULIELMUS CLEMENT. LONDIN

Month striking clock, quarter striking on three bells. 1 1/4-seconds' pendulum (shown by the side of the case). Oak case, veneered ebony. 6 ft. 2 in. high. 8 in. dial.

c. 1675

FIG. 62.

Enlarged dial of fig. 61.

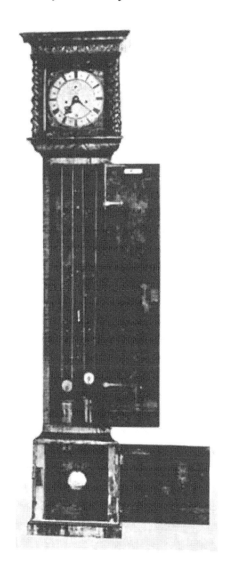

FIG. 63.

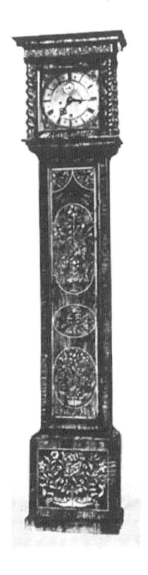

FIG. 64.FIG. 63 & 64.

"WILLIAM CLEMENT. FECIT" (c.c. 1677. Master 1694)

Eight-day striking clock. Bolt-and-shutter maintaining power. 1 1/4-seconds' (61 in.) pendulum. Case of oak, veneered walnut and inlaid floral marqueterie in panels. Lift-up hood. 6 ft. 5 3/4 in. high. Dial 10 in. wide × 10 1/8 in. high.

c. 1685

FIG. 65.

Enlarged dial of figs. 63 and 64.
THOMAS TOMPION

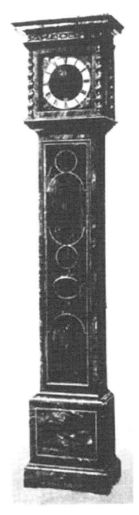

FIG. 66. "THOMAS TOMPION. LONDINI"

Thirty-hour striking clock. Lift-up hood. Key-wind. Oak case, veneered walnut and inlaid with stringing. 6 ft. 7 3/4 in. high. 10 in. dial.

c. 1675

FIG. 67.

Enlarged dial of fig. 66.

FIG. 68.

"THOMAS TOMPION. LONDINI. FECIT"

Eight-day striking clock. Bolt-and-shutter maintaining power. Lift-up hood with carved cresting. Case of oak, veneered walnut, and inlaid star marqueterie of coloured woods. 6 ft. 5 in. to top of hood. 10 in. dial.

c. 1680

FIG. 69.

Enlarged dial of fig. 68.

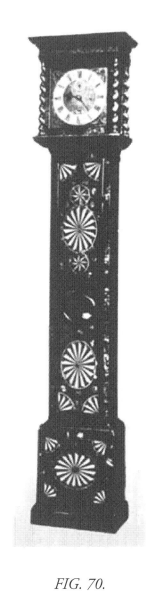

FIG. 70.

"THO; TOMPION. LONDINI. FECIT"

Month striking clock, No. 69. Bolt-and-shutter maintaining power. Case of oak, veneered walnut, and inlaid with fan pattern marqueterie in panels. 6 ft. 6 in. high. 10 in. dial.

c. 1685

FIG. 71.

Enlarged dial of fig. 70.

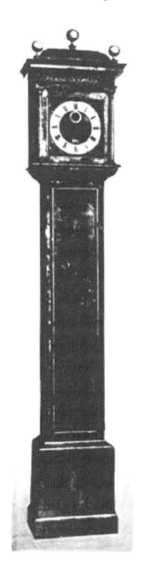

FIG. 72.

"THO; TOMPION. LONDINI. FECIT"

Eight-day striking clock. Bolt-and-shutter maintaining power. Lift-up hood. Case of oak, veneered walnut. 6 ft. 11 in. high to top of hood. 10 8/16 in. dial.

c. 1685

FIG. 73.

Enlarged dial of fig. 72.

FIG. 74.

"THO; TOMPION; LONDINI; FECIT"

Month striking clock, No. 292. Bolt-and-shutter maintaining power. Case of oak, veneered walnut. 8 ft. 6 1/4 in. total height to central spire. 12 in. dial.

c. 1690

FIG. 75.

Enlarged dial of fig. 74.

FIG. 76.

"THO; TOMPION. LONDINI. FECIT"

Month striking clock, No. 301. Bolt-and-shutter maintaining power. Case of oak, veneered walnut. 7 ft. 2 in. high to top of cornice. 11 in. dial.

c. 1695

*FIG. 77.*Enlarged dial of fig. 76.

FIG. 78.

"THO; TOMPION; LONDINI; FECIT" *(c.c. 1671, Master 1704)*

Month striking clock, No. 371. Bolt-and-shutter maintaining power. Case of oak, veneered pollarded olive wood. 7 ft. 1 3/4 in. high to top of cornice. Pull forward hood. Clock and case numbered No. 371. 11 1/8 in. dial.

c. 1695

FIG. 79. Enlarged dial of fig. 78.

THOMAS TOMPION AND EDWARD BANGER

FIG. 80.

"THO; TOMPION; EDW; BANGER" "LONDON"

Eight-day striking clock. Bolt-and-shutter maintaining power. Oak case, veneered ebony, and numbered No. 351. 6 ft. 11 in. high to cornice. 11 in. dial.

c. 1710

FIG. 81.

Enlarged dial of fig. 80.
THOMAS TOMPION

FIG. 82.

"THO. TOMPION. LONDINI"

Three-month striking clock. Engraved arch to dial. Oak case, veneered ebony, with gilt brass enrichments. 8 ft. 6 in. high. Dial 13 in. × 17 in.

c. 1710

FIG. 83.

Enlarged dial of fig. 82.

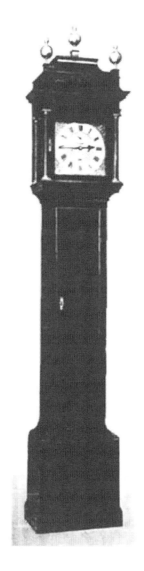

FIG. 84.

"THO; TOMPION; LONDON"

Eight-day striking clock. Bolt-and-shutter maintaining Power. Oak case, veneered ebony, and numbered No. 504. This is probably the latest long-case clock by this maker, as it is the highest recorded number. 6 ft. 9 1/4 in. high to cornice. 11 in. dial.

c. 1712

FIG. 85. Enlarged dial of fig. 84.

HENRY JONES

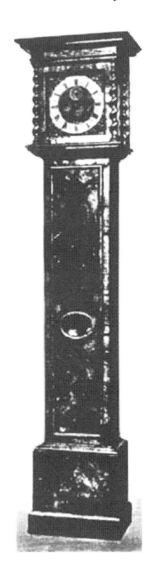

FIG. 86.

"HENRY JONES IN YE TEMPLE"

Eight-day striking clock. Bolt-and-shutter maintaining power. Case of oak, veneered walnut. Lift-up hood. 6 ft. 4 1/2 in. high. 10 in. dial.

c. 1680

FIG. 87.

Enlarged dial of fig. 86.

FIG. 88.

"HENRY JONES. LONDON"

Eight-day striking clock, bolt-and-shutter maintaining power. Case of oak, veneered walnut, and inlaid with floral marqueterie in panels. Lift-up hood; outside locking-plate; 6 ft. 4 1/2 in. high to cornice. 10 in. dial.

c. 1685

FIG. 89.

Enlarged dial of fig. 88.

FIG. 90.

"HENRY JONES IN YE TEMPLE"c.c. 1663. Master 1691

Eight-day striking clock, bolt-and-shutter maintaining power. A very early example of a dial with ringed winding holes. Case of oak, veneered walnut, and inlaid with floral marqueterie in panels. 6 ft. 6 1/2 in. high to cornice. 10 1/8 in. dial.

c. 1690

FIG. 91. Enlarged dial of fig. 90.

JOSEPH KNIBB

FIG. 92.

"IOSEPH. KNIBB. LONDINI. FECIT"

Eight-day striking clock. Striking on "Roman numeral" system (see Glossary). (Note the "IV" on dial instead of the usual "IIII.") Case of oak, veneered walnut, and inlaid floral marqueterie in panels. Lift-up hood. 6 ft. 5 in. total height. 10 in. dial.

c. 1685

FIG. 93.

Enlarged dial of fig. 92.

FIG. 94.

"IOSEPH. KNIBB. LONDINI. FECIT"

Eight-day, three-train, striking and chiming clock. Case of oak, veneered walnut, and inlaid floral marqueterie in panels. Lift-up hood. It is probable that these cases, without plinths, were made to stand in moulded adjustable trays. 6 ft. 3 1/4 in. high. 10 in. dial.

c. 1685

FIG. 95.

Enlarged dial of fig. 94.

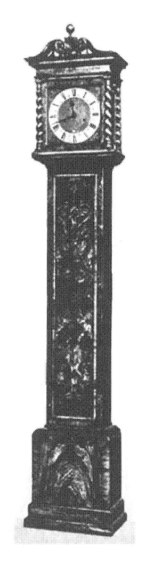

FIG. 96.

"IOSEPH KNIBB. LONDINI. FECIT"c.c. 1670

Eight-day striking clock. Lift-up hood with additional bolt. Oak case, veneered walnut. 6 ft. 4 1/8 in. high to top of cornice. 9 5/8 in. dial. (Note Joseph Knibb's square-top striking bell, which is characteristic of his clocks).

c. 1695

FIG. 97.

Enlarged dial of fig. 96.

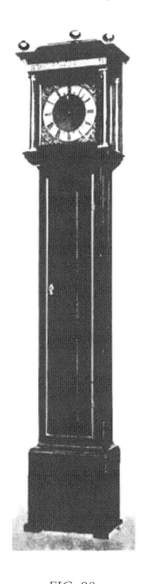

FIG. 98.

"IOSEPH KNIBB; LONDON"

Month striking clock, striking on two bells on Roman numeral system (see Glossary). Case of oak, veneered ebony. Lift-up hood. 6 ft. 6 7/8 in. high to top of cornice. 10 in. dial.

c. 1695

FIG. 99.

Enlarged dial of fig. 98.

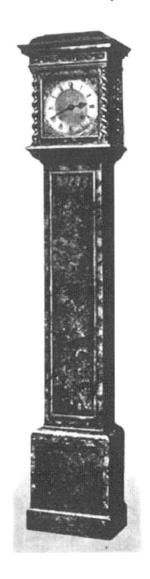

FIG. 100.

"JOSEPH KNIBB. LONDON"

Three-month striking clock, striking on two bells on Roman numeral system (see Glossary). 1 1/4 seconds pendulum. Case of oak, veneered burr walnut. 7 ft. 3 1/8 in. high. 10 5/8 in. dial.

c.c. 1705

FIG. 101.

Enlarged dial of fig. 100.
JOHN KNIBB OF OXFORD

FIG. 102.

"JOHANNES KNIBB; OXONIÆ. FECIT"

66

Eight-day striking clock. Case of oak, veneered walnut, and inlaid floral marqueterie in panels. Lift-up hood. 6 ft. 6 1/4 in. high. 10 in. dial.

c. 1695

FIG. 103.

Enlarged dial of fig. 102.
DANIEL QUARE

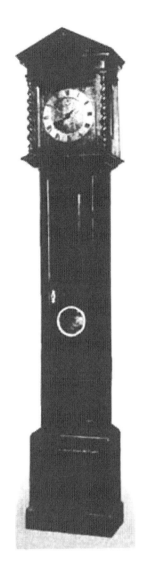

FIG. 104.

"DANIEL QUARE IN MARTINS-LE-GRAND" c.c. 1671.

30-hour striking clock; one hand. Engraved centre to dial. Outside locking-plate. Case of oak, veneered ebony. 6 ft. 11 3/4 in. high.

Portico-top with spiral pillars to hood. Lift-up hood.

c. 1675

One of Quare's very early clocks.

FIG. 105.

Enlarged dial of fig. 104.

FIG. 106.

"DANIEL QUARE. LONDON"

Eight-day striking clock. Outside locking-plate. Case of oak, veneered walnut and inlaid floral marqueterie in panels. Carved cresting to lift-up hood. 6 ft. 6 1/16 in. to top of cornice. 10 in. dial.

c. 1685

FIG. 107.

Enlarged dial of fig. 106.

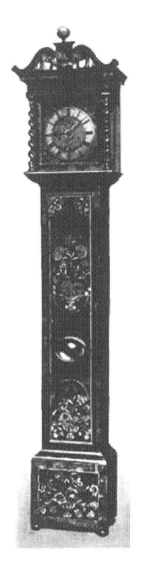

FIG. 108.

"DANIEL QUARE; LONDON"

Eight-day striking and alarum clock. Bolt-and-shutter maintaining power. Case of oak, veneered walnut and inlaid floral marqueterie in panels. Lift-up hood. Carved cresting to hood. 6 ft. 4 3/16 in. to top of cornice. Dial 10 in. Inside locking-plate. Alarum disc behind hand collet.

c. 1690

FIG. 109.

Enlarged dial of fig. 108.

FIG. 110.

"DAN; QUARE. LONDON"

Eight-day striking clock. Dial with ringed winding holes. Case of oak, veneered with walnut and laburnum "oyster pieces." 6 ft. 11 1/4 in, high to top of cornice. 12 in. dial.

c. 1710

FIG. 111.

Enlarged dial of fig. 110.

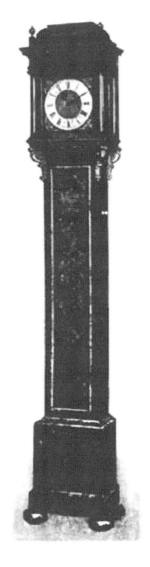

FIG. 112.

*"DAN; QUARE. LONDON"*c.c. 1671. Master 1708.c.c.

Eight-day striking clock, pull-repeater on two bells. Case of oak, veneered walnut, unusual "bun" feet for adjustment. 5 ft. 9 3/4 in. high. 7 1/2 in. dial. Carved gilt trusses under hood.

c. 1710

FIG. 113.

Enlarged dial of fig. 112.

FIG. 114.

"DANIEL QUARE; LONDON; No. 145"

Month striking clock on 6 bells (Grand sonnerie; see Glossary). Case of oak, veneered fine scrolled marqueterie, also on pillars; frets on sides and frieze (very rare at this date). 8 ft. 7 in. high to top of moulded platform. Dial 12 in. wide × 16 3/4 in. high. Day of month and strike-silent fingers in arch.

c. 1720

FIG. 115.

Enlarged dial of fig. 114.

FIG. 116.

Details of the Quare Clock, fig. 114.

FIG. 117

Enlarged view of cornice of fig. 114.

This Quare clock is instructive in many ways, perhaps to a greater degree than any other example in this book. Being of large size—it is 8 ft. 7 ins. high without the central spire and its base—it must have been made for a very tall room, as it would be utterly out of scale in an apartment of anything less than 20 feet in height. Being a three-train month striker, Grand Sonnerie on six bells (very unusual in a long-case clock at this early date—or any other), everything about the clock is special. Dan Quare numbered royalty among his clients, as there are clocks by him at Hampton Court and Buckingham Palace and elsewhere, and this may have been a Palace clock originally. The planting of the dial has necessitated six spandrel corners, and these are of individual design, not the products of the usual type of "clock-parts" foundry.

It is the case itself which is highly unusual. It is of the arch-dial form (by no means an innovation in 1720, as we have the Tompion clock in the Pump Room at Bath, and its fellow at Iscoyd Park, and at least one other example illustrated in this book, all of somewhat earlier date) but it is inlaid with marqueterie of fine scrolled pattern of holly in a ground of walnut, which was almost an anachronism in 1720, as this type of marqueterie is one of the latest phases of the inlaid fashion. This vogue was comparatively short-lived, especially in furniture, and with the introduction of the arch-dial case (which only became general after about 1725, if not later), the taste for marqueterie had declined in favour of plain walnut of the burr or other figured kinds. True, while arch-dial marqueterie cases are rare (they are not unknown; three examples are illustrated on page 91); one has always the suspicion that they are altered clocks, not made as recent forgeries but by using up stock models, adding the arches to the existing dials, or by other means. This cannot be said about this Quare clock; it is original throughout, just as it was made in circa 1720, obviously to a special order from an important source.

FIG. 118.

Enlarged trunk door of fig. 114.

There was an important reason why marqueterie would tend to go out of fashion, one dictated as much by the makers themselves as by their clients. Marqueterie is the inlay of one, or more, veneers in another—the ground. Cut by hand, with a fine saw in a large frame, working in guides, it is impossible to cut in any less than four layers, plus outside rough veneers to take up the "rag" of the saw. Even with this duplication, there is the counterpart (inlay of dark wood in light ground, the "fall-out" of the original cutting, which would not be thrown away, as a rule), that has to be considered. This means that with a clockcase, such as the one we are considering here, there is an overplus inlay for seven more, three like the first, and four in reverse. Panels such as in Fig. 118 are frequently cut in two halves or four quarters as an economy, but that is not the case here. The panel of this trunk door is in two vertical halves only. Duplicates of the same inlay must have existed, therefore (three at least if we reckon counterparts, and everywhere the work cut in balancing halves), and the early makers were against making several cases of identical patterns. In this clockcase we have, also, the very early use of fretwork, here cut with the marqueterie saw, a very difficult method, as the saw has to be detached with each complete hole in the pattern, and the marqueterie saw does not remove as easily as the later fret saw, and, also, the pattern has to be cut vertically, instead of horizontally, as with the fret saw.

EDWARD ORTON

FIG. 119.

EDWARD ORTON, LONDON c.c. 1687

Eight-day striking clock with apertures in dial for day of month, signs of the Zodiac, etc. Case of oak, veneered walnut, and inlaid floral marqueterie in panels. 7 ft. 3 in. high to cornice. 11 1/8 in. dial.

c. 1705

FIG. 120.

Enlarged dial of fig. 119.
GEORGE GRAHAM

FIG. 121.

"GEO; GRAHAM. LONDON"

Eight-day striking clock. Bolt-and-shutter maintaining power. Dead-beat escapement. Dial engraved in two places. Case of oak, veneered pear wood, and ebonised. 7 ft. 2 in. high. 11 in. dial.

c. 1720

FIG. 122.

Enlarged dial of fig. 121.

FIG. 123.

*"GEORGE GRAHAM. LONDON"*c.c. 1695. Master 1722

Eight-day striking and equation clock. Case of oak, veneered walnut. 7 ft. 9 1/2 in. high. Dial 12 in. × 16 1/2 in.

c. 1730

FIG. 124.

Enlarged dial of fig. 123.

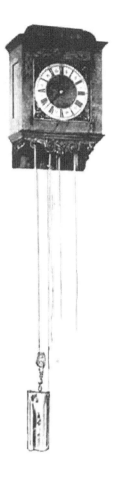

FIG. 125.

"CHRISTOPHER GOULD. LONDON"

Eight-day striking wall clock, pull-repeating on two bells.
Lift-up hood. Walnut case. 6 1/8 in. dial. *c. 1695*

91

FIG. 126.

"JOHN KNIBB. OXON. FECIT"

Pull-up wind, striking wall clock. Walnut case. Bob pendulum. Lift-up hood. Outside locking-plate. 12 in. dial.

c. 1685

FIG. 127.

FIG. 128.

FIGS. 127 & 128.

"THO; TOMPION. LONDON"

30-hour striking and alarum wall clock, in walnut case. Pull-up winding.

c. 1700

FIG. 129.

"GEORGE GRAHAM. LONDON"

30-hour alarum wall clock. Mahogany case (possibly a later addition). Pull-up winding.

c. 1725

LATE ARCH-DIAL CLOCKS IN WALNUT CASES

FIG. 130.

JOHN ELLICOTT LONDON

1706—1772

Eight-day striking clock. Arch dial with strike-silent in arch. Case of oak, veneered burr-walnut. Dial 12 in. × 16 in. 7 ft. 6 in. high without spire.

c. 1740

FIG. 131.

JOHN ELLICOTT LONDON

Month astronomical clock with circular-date calendar in trunk door, showing dates before and after alteration of the calendar in 1752. Dial divided I to XII repeated twice. Case of oak, veneered walnut. 8 ft. 10 1/2 in. total height.

c. 1760.

SQUARE-DIAL CLOCKS IN MARQUETERIE CASES

FIG. 132. c. 1675

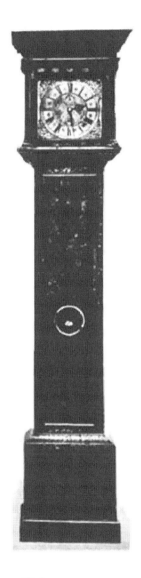

FIG. 133. c. 1705

FIG. 134. c. 1710

FIG. 135. c. 1715

FIG. 136. c. 1715

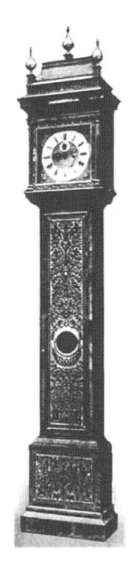

FIG. 137. c. 1720

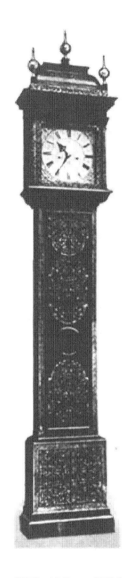

FIG. 138. c. 1710

FIG. 139. c. 1710

FIG. 140. c. 1710

Square-dial clocks with superstructures to hoods. Cases in floral marqueterie in panels on walnut ground.

ARCH-DIAL CLOCKS IN MARQUETERIE CASES

FIG. 141. c. 1720

FIG. 142. c. 1720

FIG. 143. c. 1725

Arch-dial clocks in marqueterie cases are very rare as, when the arch became general, the fashion for marqueterie had gone out.

111

LATE ARCH-DIAL CLOCKS IN WALNUT CASES

FIG. 144.

MOSES FONTAINE & D(ANIEL) TORIN

Eight-day striking clock. Phases of the moon in arch of dial. Oak case, veneered walnut. 8 ft. 3 in. high without spire. Dial 12 in. × 16 1/2 in.

FIG. 145.

JOSEPH BORRELLI LONDON

LONG-CASE CLOCKS IN LACQUERED CASES

FIG. 146.

CHRISTOPHER GOULD LONDON

c.c. 1682

Eight-day striking clock. Bolt-and-shutter maintaining power. Finely pierced hands and central boss; numbered minutes. Case of oak, lacquered on black ground. 7 ft. 8 in. high. 12 in. dial.

c. 1695

FIG. 147.

GEORGE ALLETT LONDON

c.c. 1691

Eight-day striking clock. Bolt-and-shutter maintaining power. Case of oak lacquered on a ground of powder blue. Silver mounts and spires. 7 ft. 5 in. high (without spire). 11 in. dial.

c. 1705

FIG. 148.

~

DAVID (DIEGO) EVANS ROYAL EXCHANGE

Eight-day striking and musical clock, made for the Spanish market. Case of gold ornament on a ground of red lacquer. 8 ft. 2 in. high without spire. Dial 12 in. × 16 1/2 in. high.

c. 1770

FIG. 149.

JAMES MARKWICK LONDON

c.c. 1692. Master 1720

Month striking and chiming clock. Case of oak lacquered on a blue ground. 9 ft. 3 1/2 in. total height. Dial 13 1/2 in. × 19 1/2 in. high. Carved gilt spires and trusses under hood.

c. 1725

PRECISION CLOCKS WITH FINE MOVEMENTS

FIG. 150.

Thomas Mudge

Mahogany case.c. 1785

FIG. 151.

Thomas Mudge 1715-1794

FIG. 152.

William Dutton 1746-1794

Walnut case.

124

FIG. 153.

JOHN ARNOLD, LONDON

Month Regulator Clock. (No. 1.) 24-hour indicator in lunette below hands, seconds in centre, outer circle seconds and minutes. Bolt-and - shutter maintaining power, compensated pendulum. Dial 12 1/8 in. Case of mahogany, 6 ft. 4 in. high.

c. 1784

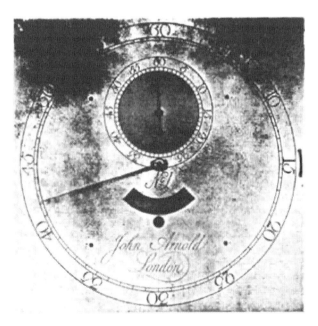

FIG. 154. Enlarged dial of fig. 153.

TYPES OF CLOCKS IN MAHOGANY CASES
OF 1775-90

FIG. 155

FIG. 156.

FIG. 157.

MAHOGANY LONG-CASE CLOCKS MADE FOR SHOP

FIG. 158.

FIG. 159.

FIG. 160.

LANCASHIRE AND YORKSHIRE CLOCKCASES

FIG. 161.

FIG. 162.

FIG. 163.

THOMAS CHIPPENDALE'S DESIGNS FOR CLOCKCASES

FIG. 164.

FIG. 165.

FIGS. 164 & 165.

From the 1st (1754) edition of the "Gentleman and Cabinet Maker's Director."

137

FIG. 166.

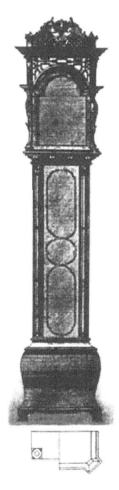

FIG. 167.

FIGS. 166 & 167.

From the 1st (1754) edition of the "Gentleman and Cabinet Maker's Director."

THOMAS SHERATON'S DESIGNS FOR CLOCKCASES

FIG. 168.

FIG. 169.

FAMOUS CLOCKMAKERS

GEORGE GRAHAM
1673-1751

THOMAS MUDGE
1715-94

JOHN ARNOLD

1736-99

JOHN ELLICOTT

1706-72

THE COMPANIONABLE GRANDFATHER

ALL CLOCKS TICK, but a grandfather 'talks', and he 'talks' impressively. His small brother, the bracket, tends to clip his words and hurry his conversation while the later generation of the clock family with a lever movement gabble unintelligibly. In the room where this is being written, there is a grandfather whose slow soft voice is undisturbed by the chatter of the typewriter; and always companionable, he is never obvious yet invariably his 'words' blend with the mood of the moment.

When the mind begins to tire and thoughts and ideas are restricted, he seems to say, 'Don't press', 'Have a rest'; or if the sunshine pulls hard against work, that same old clock will admonish, 'Don't play', 'Get busy'. At the start of the day he stimulates, and in the evening when the curtains are drawn and you lounge in your pet chair, he says softly 'For-get', 'Re-lax'; and with his quietly persistent, but never monotonous, murmuring he brings a sense of well-being and pleasant detachment from the modern hurly-burly which is born of the thing we call progress.

He is not of aristocratic lineage, nor has he a long line of ancestors, but he represents all the various features which illustrate the evolution of his kind from the early lantern clocks which have been described in Chapter Six. A grandfather clock is to be found in many homes valued for its association with past generations of the family to whom it has come over the years. Often, too, that same grandfather stands silent and moribund for lack of some slight adjustment and so of little actual use other than to introduce a certain romance of the long-ago.

Nor is that the only romantic aspect associated with some of these time-tellers even if the additional 'romance' is of a material character. Many instances might be mentioned where a grandfather clock, which 'has been in the family for years' and accepted merely as part of the household furniture or perhaps neglected and allowed to remain silent, has eventually very appreciably increased the owner's liquid assets.

Here is one such 'find' with which the writer was connected—even if he were not the 'fortunate': He was asked by a mutual friend to look over some furniture belonging to a lady and decide which was worth keeping and which should be disposed of. There were several fine pieces in the house which, having been familiar with them since childhood, the owner regarded as of little value and had in fact put aside as unworthy of notice—several of these did not find their way to a country auction room but are now highly respected and, incidentally, equally valued.

In the corner of one room, unhonoured and almost concealed by other furniture stood a small clock. The very sight of it meant recognition and a warm glow, but to the owner it was, 'An old clock that has been in the family for years. It goes, but it's not worth much.' When this writer mentioned its approximate value, he was doubtless regarded as a candidate for a home where weak-minded people are cared for—later, when she had been convinced, the anxiety displayed by the owner proved how true it is that from ignorance our comfort flows. Previously, it was merely a clock that told the time; then suddenly it took the form of the wealth of Midas which had to be guarded and cared for. The first reaction to this was that the clock must not be sold, but shortly afterwards it changed hands for a sum well in excess of that originally suggested.

In describing this horological aristocrat, we purposely omit the name of the maker, for the clock is such a rarity that it could be identified with its present owner. It is only slightly over five feet tall with a small square dial which places it in the category of the so-called 'grandmother' clocks—a name traditionally applied on the assumption that grandmother, not being as tall as grandfather, could not wind the taller grandfather clocks, but had no difficulty with the shorter ones (XXIV).

Interesting as the small size of this clock may be, its real importance is in the movement for it is one of the earliest known examples of a long case which repeats the quarter hours. A light cord hangs from the side of the case and when this is pulled the quarter hours are repeated on five bells. A similar clock by the same maker was formerly in the Wetherfield collection, a fact which helped materially in the quick identification of its 'twin' in the house where it had long been unrecognized.

After which digression, we may return to the more plebeian grandfather that is recording the passing seconds in this room; while it shows nothing that would at first sight suggest anything of the earlier lantern clock, we may, by 'dissecting' it, trace its descent from that branch of the clock family.

As explained in Chapter Six and illustrated in (16), the frame of the lantern clock consisted of a top and bottom plate joined by four turned brass pillars with a brass door at each side, a plate at the back and the dial plate in front. Only two brass plates are used with a grandfather movement, one in front and one at the back which are connected by a stout turned brass horizontal pillar at each corner. The first movements, like the later lantern clocks, had the short pendulum with a small bob, the dial being square and quite small—about 8 inches.

As might be expected, examples of this early type are not numerous and while one of them is, at first sight, similar to later grandfather clocks, two differences are obvious, i.e. the square dial is much smaller and the case is quite narrow. The last feature is explained by the fact that the short pendulum swung in the hood of the case, consequently the case needed to be only wide enough to accommodate the weights. When, however, the long pendulum was adopted, this extended down and swung in the waist or trunk which necessitated the case being slightly wider.

It would seem that the first progression toward the all wooden case was the enclosure of a brass lantern clock in a wooden hood which was supported on a bracket with the weights hanging below. And it would be natural that this use of wood to cover the movement would suggest that the weights which were by no means ornamental could be similarly enclosed in what amounted to a panelled cupboard and so the first form of grandfather clock made its appearance (19).

When the anchor escapement (described in Chapter Five) and the long or royal pendulum were adopted, the grandfather clock, as we moderns know it, became the standard time-teller in the homes of this country. In running order any one of these clocks is a reliable timekeeper and, if at any time it becomes slightly inaccurate, this can be easily remedied by adjusting the pendulum bob. The bottom end of the pendulum rod has a screw thread with a small nut; should the clock tend to lose time, this can be checked by screwing up the nut and so raising the bob, contrawise if the clock gains, the nut should be screwed downward thus lowering the bob. The amount of raising or lowering should be very slight and by watching the clock for a few days the exact degree can soon be ascertained.

As that now commonplace object, the pendulum, is very little understood or appreciated, we might here range for a short spell into the fields of explanation: If you look at the pointer in the small dial, usually in the upper part within the hour ring of a grandfather clock, you will see that it jumps one division or one second and then recoils slightly—the word 'second' meaning the one that follows or comes after. To ensure that each swing shall record one second, it was discovered that the pendulum had to measure 39.13 inches from the point where it bends the small piece of spring by which it is suspended and the point of gravity of the whole pendulum which is near to, but somewhat higher than, the centre of the disc or bob.

Therefore the pendulum of a clock of the usual grandfather type must make 86,400 swings to record the time accurately for the twenty-four hours we call a day. If the clock, at the end of that time, has gained, say, six minutes, it shows that the pendulum has made 360 swings too many or if it has lost that number of minutes the pendulum has made 360 swings too few. And each of these faults is remedied by the small nut under the bob by which the centre of gravity is varied—in plain language, the pendulum is lengthened to lessen the number of swings and check the 'gaining' or shortened to increase the number of swings to check the 'losing'.

One of our modern truisms is that demand controls the monetary value of an article and this applies to grandfather clocks with thirty-hour movements. Admittedly, it is slightly more trouble to have to wind a clock every day instead of once a week, but that does not detract from its ornamental qualities; and many of the thirty-hour grandfathers are as attractive as their more appreciated eight-day fellows.

When clocks were first enclosed in long wooden cases most of them had thirty-hour movements which were driven by weights on a strong cord. Many were wound in the same way as the lantern clocks, namely by pulling down the cords and raising the weights and these may be recognized from the absence of winding holes in the dial which are found with clocks which are wound up by a key. When, however, the going and striking trains were placed side by side instead of one behind the other, the thirty-hour grandfathers could be and were wound by a key and the dials of these have the winding holes.

It is commonly thought that the thirty-hour clocks were made only by the less experienced provincial clockmakers. Admittedly, the larger number of those which are still to be found in country houses, cottages and inns were the work of some local man, but examples were made by, now famous, London clock-makers. One by the celebrated

Thomas Tompion in the Guildhall Museum has a lantern movement which is wound by pulling down the cords and raising the weights; another, by the same maker, which was in the Wetherfield collection, has the two winding holes in the dial.

After the long pendulum was introduced, it might almost seem that the more prominent makers vied with each other in producing a movement which would run for the longest period between windings. It is generally supposed that the maximum period during which a clock will run without winding is eight days. With very few known exceptions, this is true with spring-driven clocks, but as a result of experiments based upon close calculations several of the eighteenth-century clockmakers produced grandfather clocks which would continue going for a whole month and some are known which will run for three, six and twelve months. Any which will run for the last three mentioned periods are rare, but there are an appreciable number of month duration—there were upward of thirty in that part of the Wetherfield collection which went to America.

To achieve these longer going periods called for more intricate mechanism, a larger number of wheels and stronger driving power. This additional drive could be obtained with a weight-driven clock by using heavier weights, but it was impracticable with a spring-driven clock both from the point of appearance and the prohibitive cost.

While not a grandfather, a clock which strikes the hours and quarters and runs for a year without winding, illustrated by F. J. Britten, will serve to show how costly such a curiosity could be. It was made for William III by Thomas Tompion and is said to have been in the king's bedroom at Kensington Palace when he died. It was bequeathed to the Earl of Leicester and now belongs to Lord Mostyn whose family have owned it for some two centuries.

Tompion's bill for making the clock was £1,500 which has to be multiplied several times to arrive at the present day equivalent. The case which is of ebony with silver mounts, is in two box-like sections, the upper and smaller of which is similar to a grandfather hood with a dial, 10 inches square.

The lower section might be called the 'engine house', for it contains the two large driving wheels and the two powerful springs in the barrels with the accompanying fusees; and the many other wheels and pinions and the verge escapement connected with the striking and repeating mechanism are housed in other parts of this unusual clock.

LONG CASE CLOCKS

WHEN we examine the history of the long case clock we are confronted with a very big subject, and it is impossible in this volume to deal fully with the cabinetmaker's side. The evolution of cases by successive generations of London cabinet-makers, has been fully dealt with by Cescinsky and Webster in their remarkable work entitled "English Domestic Clocks," and every one who handles old clocks would do well to purchase a copy of this most interesting book. They say, however, practically nothing about North Country or Scottish styles. For the purpose, for which this volume is written, it will suffice if only brief reference is made to the more outstanding types.

We have already noted that lantern clocks were not infrequently "boxed in" with a view to rendering them more dust-proof, and the boxing-in of the weights and pendulum naturally led to the construction of long case clocks; but before entering into details a few remarks on various influences may not be out of place.

Between the years 1650 and 1750, London was the centre of the clockmaking industry in the same way that Sheffield has for a long time been the centre of the cutlery industry. This fact is chiefly attributable to London being the only large town at that time, and that it presented unequalled advantages for the sale of the clocks produced. Edinburgh, on the other hand, also proved a good centre, but to a lesser degree. Britten and Smith both mention the names of several first-class men who migrated to London and Edinburgh for business purposes just as rising members of the legal profession not infrequently do to-day. The centering of the industry in London had a remarkably good effect on the early products. Not only were the makers situated in a good market, but they were surrounded by many great cabinet-makers of marvellous skill and excellent taste. At that time London attracted great numbers of continental workmen, and some of these must have added considerably to both the clockmakers' and the cabinetmakers' wealth of ideas, and proved of great assistance in carrying out certain designs. The trade having once been established in London and Edinburgh, makers from other parts of the country visited these towns to procure ornamental castings and other material, but by the middle of the eighteenth century we find them producing new designs of their own which are not found on London-made clocks.

There is no doubt that the early London-made clocks left nothing to be desired in workmanship and proportions. The former was due, partly to the great pride the producers took in their work, and partly because the cost was of minor importance. Another factor, however, contributed to the maintenance of a high standard of quality, and that was the powerful influence of the Clock Makers' Company. This Guild possessed very great powers, and they could not only seize and destroy any second-rate article, but could ruin any man who did not consistently turn out first-rate work.

When, however, the powers of this great Guild waned, a reaction set in and inferior goods were produced. The growing demand for clocks naturally led to another change. At first they were only purchased by the extra wealthy, but later on other people with more slender purses required them, and the natural result was that clocks and clock cases had to be produced at a much lower figure, and many refinements had to go. It is interesting to compare the very high standard, not only of clocks, but of all sorts of beautiful old furniture in some ancient mansion with the corresponding clocks and furniture handed down from generation to generation of, say, farmers who were less prosperous. In the one case we find the choicest materials, design and workmanship, and in the other case the village craftsman's attempt to produce something of the same sort from a poor class of material, the proportions and workmanship frequently being likewise somewhat poor.

It is interesting to note that at the time the London makers were distinctly degenerating, those of Lancashire and certain other districts were producing work of great excellence. With the general slump which occurred in the clock trade owing to foreign competition, most of the provincial makers died out, and apart from movements made in factories in Birmingham, London, etc., comparatively few are made in this country at all. There still is a colony of chamber workers in London, but they are getting fewer in number every year.

Before the war the Germans were doing their utmost to stamp out the last traces of clockmaking in this country. Not many years ago the writer was shown, in a wholesale house, a sample of a finished, silvered and lacquered dial for a long case clock delivered in Glasgow for *ninepence* more than the bare cost of the brass.

The earliest long case clocks produced had certain features in common with the lantern. Many had crown escapements and short pendulums, which accounts for the fact that the early cases were very narrow. The dials were square and not more than 9 or 10 inches across. The dial centres were frequently engraved with the lily pattern so popular at this period. The corner ornaments were simple, small and beautifully finished with the graver. The hand was of the simple double loop type, frequently bevelled on the surface. The movement was of the lantern type, and when wound with a key, the two trains were placed side by side. The hood of the case had twisted pillars on each side, and in the earliest examples the hood had to be drawn off or raised before the hand could be touched. The moulding under the hood was convex instead of concave, and the same applies to the top of the plinth. The door was long, narrow and square-topped, and the distance from the floor to the top of the plinth was frequently distinctly less than in later types. The whole design was extremely simple and dignified, the only ornamentation being a very small quantity of inlay.

The reader is cautioned against regarding any one or two of the above characteristics as pointing to an early clock. If the movement is not an old one, the style of engraving of the dial and the corner ornaments are perhaps some of the best guides, providing that the dial is a genuine old one, and not merely a reproduction. On the other hand, the size of the dial, the one hand, the thirty-hour lantern movement, the plain rectangular door, the simple case with a low plinth and several other characteristics are frequently found in clocks even as late as 1760 The matter is further complicated by dealers interchanging dials and cases so that good cases may have good dials and movements and *vice versâ*. When this is done dealers ignore, as a rule, the question of age and locality.

PLATE XI.

HENRY HESTER, WESTMINSTER. ABOUT 1695.

PLATE XII.

DIAL OF CLOCK, BY HENRY HESTER.

About 1690, or perhaps even earlier, several very marked changes took place. The eight-day movement was developed; clocks with minute hands and calendar circles were introduced, and the ornamentation of cases was further developed. This period also witnessed the introduction of frosted dial centres, which became the standard practice in London for about three-quarters of a century. The clock by Hester, shown in Plates XI. and XII., dates from about 1695.

So long as the crown escapement was used in long case clocks, the aperture for showing the day of the month was located between the centre of the dial and the figure XII., but the application of the seconds pendulum and seconds hand resulted in the aperture being placed just above the figure VI. To-day the calendar on a clock is regarded as quite unnecessary, but from about 1690 to 1790 hardly any clocks, except lanterns, were constructed without them; in fact, the calendar was regarded as far more important than the minute hand. To-day this appears almost incredible, but it must be remembered that 200 years ago no one had a train to catch, and the general public had no printed calendars or daily newspapers to refer to when they wished to date their letters.

The introduction of the minute hand, of course, brought with it minute divisions on the dial. In some cases we find every minute separately numbered round the dial, but this soon gave way to the practice of numbering every fifth minute. Many early makers made their minutes very large and clear, as illustrated in Plate XII., but these were soon followed by the minute figures being placed outside the divisions, as shown in Plate XIII. The size of the minute figures tended to grow, as will be seen by examining a series of dials of different ages. A few makers in the middle of the eighteenth century even went so far as to make the minute figures as large as the hour figures, as shown in Plate XXI.

About 1700–10 we find most long case clocks fitted with dials similar to those illustrated in Plate XIII. These may be regarded as characteristic of this period, and are not only found on British clocks with marquetry cases, but also on clocks of Dutch origin. It will be noticed that an early maker's name is followed by the Latin word "*fecit*" (made it), and he usually engraved his name on the base plate below the circle. The centre is frosted, with little or no ornamentation with the exception of a Tudor rose in the middle. This was probably suggested by the alarm plates of the earlier lantern clocks (see Plate V.). Surrounding the circle will be noticed the ever-popular lily design. The corner ornaments have undergone certain elaboration. About this time hour-hands became very ornate, and many of the best makers developed some extremely intricate and beautiful designs. Some were not only elaborately pierced, but the surface was carved or bevelled in addition. Further examples of the hands of this period are given on p. 117. This period witnessed the introduction of turned circles round the winding holes. It was probably done for a dual purpose, partly for ornamentation, and partly because a frosted dial is so easily marked by the winding key. The herring-bone border was introduced about this time.

The plate-frame movement was, of course, thoroughly established for eight-day work. Outside locking plates were common, but many were constructed with inside locking plates and strike main wheel of seventy-eight teeth. The earlier clocks with rack-striking mechanism had the rack placed inside the movement. The upper pinions in the trains were made with six or seven leaves only, the fly pinions being small and light for high speeds.

ELEVEN-INCH DIAL, BY EDWARD BIRD, LONDON.
ABOUT 1700.

PLATE XIII.

TEN-INCH DIAL, BY EDWARD COCKEY, WAR-MIN-
STER. STYLE ABOUT 1710.

PLATE XIV.

ELEVEN-INCH DIAL, BY WM. CARTER, CAMBRIDGE.
ABOUT 1715.

TWELVE-INCH ARCHED DIAL, BY GEO. MAYNARD,
LONDON. ABOUT 1715

PLATE XV.

(a)

9 1/2-INCH LANDSCAPE DIAL. ROBERT BIRD, YELD-HAM.

(b)

TWELVE-INCH ARCHED DIAL. (ANONYMOUS.)

About 1715 we find the arched top added to the dial. It was probably suggested by some one making a long case to accommodate a lantern clock without hiding the fret above the dial, for we meet with such clocks occasionally. For many years the arch was used merely for appearance and not put to any useful purpose. With the introduction of the arch, we find flower or scroll engraving on the dial base just outside the circle disappearing. Additional ornamentation on a very restricted scale appeared in the centre, but makers omitted

170

the Tudor rose. Owing to the state of politics at the time, it was probably considered that the presence of a Tudor rose might seriously restrict the market for the clock. The former types of corner ornaments disappeared, and fresh designs came on the market. For the arch, the domed name-plate and dolphin ornaments shown in Plate XXI. had a very long run of popularity, but arch ornaments matching those in the corners are by no means uncommon, as shown in Plate XV. (*a*), which is only one of many examples.

A certain number of makers adopted the arch ornament shown in Plate XIV., using the centre space as a name-plate, or engraving a face upon it. Some of these ornaments have a smaller crown at the top.

No very striking changes in dials or movements took place between 1710 and 1730. Many new designs of corner ornaments were produced, but the dial centres remained much the same. Hands became standardised in design, but varied in proportions, as will be noticed from Plates XXXI. and XXXII. Cases, on the other hand, underwent a good deal of change not only in material and style of finish, but also in the design of the hood. In some examples the hood followed the contour of the arch dial fairly closely, as in Plate XXII., but others carried a more or less elaborate superstructure. The latter appear to have been suggested by the tops of bracket clocks made twenty or thirty years previously and illustrated in Plates XXV. and XXVII. Such adapted designs were for the most part quite good, especially in England. By about 1730 we find the arch being used for some mechanical contrivance such as a strike silent hand; this was soon followed by various forms of automata such as a swinging figure of Father Time (Plate XV.), a ship in full sail (Plate XXIV.), or some mechanism for showing the phases of the moon (Plate XVII.). Other changes in the dial are noticeable. The corner ornaments became distinctly degenerate, not only as regards design, but also in workmanship, as will be seen from Plate XV. (*a*). The quarter hour divisions on the circle were omitted, the hour hand was made longer, and of less pleasing design. By 1730 most of the movements had outside racks, but the tail of the gathering pallet had not come into general use. During the period under consideration we find that clockmaking outside London and Edinburgh was becoming a very important industry, especially in Lancashire, a county which for many years held a high reputation not only for clocks but for all sorts of clockmaker's tools. Even to-day one frequently sees certain tools in catalogues described as *Lancashire pattern*. While London clocks were degenerating, the Lancashire men were rapidly improving, and so far as mahogany cased clocks were concerned the north countrymen frequently excelled the Londoners in their products.

PLATE XVI.

MAHOGANY-CASED CLOCK, BY JOHN WYKE, LI-VARPOOL.

PLATE XVII.

THIRTEEN-INCH DIAL, BY JOHN WYKE, LIVARPOOL.

PLATE XVIII.

LANCASHIRE OAK CASE.

Plates XVI. and XVII. show a mahogany clock and dial by John Wyke, Livarpool. The case is one typical of the district, though many were also made with double columns at each side of the hood. The general outline of the hood is typical of the mahogany ones of the district and should be carefully noted. The base of the case is somewhat unusual, but by no means unique, since many were made so in Lancashire and Cheshire. Probably one of the most remarkable features about this case is the immense amount of detail in all the mouldings.

In the old London-made clocks we frequently find fretwork openings backed with silk to allow the sound of the bell to escape, and afford some touch of ornamentation. Some Lancashire men followed their example, and subsequently developed the practice of introducing pieces of glass into their hoods and ornamenting them with some gold design on a dark blue background. In this clock by John Wyke, the ornamental glass has been damaged and painted over. The somewhat elaborate platform between the horns of the hood is peculiar to the district. In many cases this platform was surmounted by a large fluted ball and spike of wood. The dial is a very fine example of Lancashire work, but the second hand is not original.

Probably in no district did the oak case attain such a high standard of perfection as in the north-west counties of England. A typical example is shown in Plate XVIII. Like many others made in this district, the case is provided with bands of mahogany round the door and the panel of the plinth. A certain amount is incorporated in the hood and elsewhere. The design of the case throughout shows that the northern maker had a good eye for proportions. The lines of the hood bear a great contrast to those made in some other parts of the country at this period and which were surmounted by a somewhat heavy looking superstructure, especially in Scotland. They compare very favourably with the crudely jagged tops so frequent in other districts.

The north countryman displayed more enterprise and imagination than his London contemporary as regards dial making. While the Londoner was getting slack and disinterested in brass dials, the Lancashire men were developing new designs and turning out first-class workmanship. They frequently adopted London types of corner ornament, as will be noticed in Plate XIX. (*a*). In fact, this dial was probably made twenty-five years after this particular corner ornament disappeared from London workshops. Very soon they developed new designs, just as men in other districts did. After 1720 we find not only a great sameness about the centres of London dials, but many of the corner ornaments, if not actually degenerate, were poor in quality, being spoilt by fins appearing in the open work and by sand marks on the surface. On the other hand, the north countryman almost invariably filed away all rags and fins produced in casting and took care that their ornaments had good surfaces, free from sand marks, before they attached them to dials.

Some of the dial ornaments found on old clocks are illustrated by Cescinski and Webster, but a complete catalogue has never been published. In the case of the very earliest corner ornaments it is easy to arrange them in chronological order, but later on great complications arise. We have already seen that long after London makers had discarded some of the better designs, provincial makers were adopting them. The whole subject requires very careful tabulation and analysis, and perhaps some one will undertake this work. One is inclined to regard the set of cast corners representing the seasons of the year as comparatively recent, since we find it on dials from 1750 onwards. It will be observed that the engraved corners of the dial shown in Plate XXIV. (*a*) are of this type. On the other hand, this design is really an old one, for it is found on a sixteenth century stone dial in Exeter and on the Stuart clock illustrated in Plate VI. Sometimes makers attached engraved corner ornaments to their dials, especially in Scotland.

(a)

TWELVE-INCH FROSTED AND ENGRAVED DIAL, BY
F. KNOWLES, BOLTON.

PLATE XIX.

(b)

THIRTEEN-INCH HERALDIC DIAL. BARNISH, ROCH-DALE.

PLATE XX.

(a)

TEN-INCH "IMARI" DIAL. WM. KENT, WALDEN.

(b)

TEN-INCH "NANKIN" DIAL. EDW. SAMM, LINTON.

Provincial makers and north countrymen in particular did much to relieve the monotony of the frosted dial centre. It is probable that they did far more in this respect than the London men. In the neighbourhood of London we find that engraved centres were chiefly ornamented with rather conventional scroll designs. In fact, many engravers appear to have been incapable of producing any new ideas in this respect. Among provincial makers we find attempts to break away from standard forms. Thus, in the centre of the dial by William Kent of Walden (Saffron Walden, Essex) (Plate

XX.), we find foliage, birds and butterflies. Not only this, but part of the engraving is filled with red wax, with a rather good effect. It is evident that a piece of Imari china suggested the design and red waxing. The bird and butterflies are very similar to those found on Oriental china from Japan.

The dial by Edward Samm, Linton (near Cambridge) (Plate XX.), is another example of an attempt to adapt an Oriental design, but with a very poor result. The engraver appears to have realised that it might not be recognised as a Nankin design, so has inserted a diminutive Chinaman between the centre and "four o'clock."

The dial by Barnish of Rochdale (Plate XIX. (*b*)) has an heraldic design engraved in the centre; this was probably done to order, since it represents the arms of a Prime Bailiff of the Weaver's Company. The circle of this dial has the minutes arranged in waves, a thing frequently seen in old watches, but not very often in clocks.

The dial already referred to in Plate XIX. (*a*), is a splendid illustration of what can be done by a combination of frosting and engraving, and the dial shown in Plate XXI. (Alex. Rae) is another good example. Both Lancashire and Scottish makers frequently used sunk seconds, as shown in the Essex landscape dial illustrated in Plate XV. (*a*), an admirable plan, which saved all risk of the second hand catching on the hour hand. In the case of the dial shown in

Plate XXI. the silvered seconds circle is "inlaid" in the base plate.

In the southern counties of England we usually find quarter-hour. divisions only for one hand clocks, but makers in the north very frequently divided the space between the hours into twelve spaces (five minutes each). Sometimes these divisions were near the inner edge of the circle and sometimes near the outer edge, where the minute divisions are usually placed.

A departure from the usual practice is frequently found in Lancashire and Cheshire dials, viz., the indication of the minutes on the circle by a series of round dots, at the same time omitting the two lines one usually associates with minute divisions. London makers, apparently, only applied moon discs to arched dials, but provincial makers frequently applied them to square dials. Frequently this was done by means of a circular opening as shown in Plate XIX. (*b*), but an opening of shape similar to that in the arch was also quite common.

TWELVE-INCH DIAL, FROSTED AND ENGRAVED
CENTRE AND BAGPIPE CORNERS. ALEXr. RAE,
DUMFRIES. ABOUT 1760.

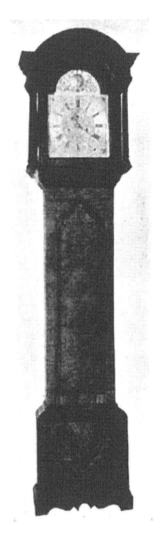

PLATE XXII.

WALNUT-CASED CLOCK, BY JOHN BROWN, LEVER-POOL.

TWELVE-INCH DIAL, BY JOHN BROWN, LEVERPOOL.

Although two hand eight-day clocks were made in large numbers before the end of the seventeenth century, it is remarkable that the making of thirty-hour and one hand clocks should have persisted for so many years. The addition of the minute hand could not have materially added to the cost. Certainly many of the thirty-hour clocks were somewhat roughly made, but many others were of really high-class. construction, showing that the maker would not have had the slightest difficulty in producing a first-class eight-day movement. At the same time they very frequently provided sham winding holes and squares. Plates XXII. and XXIII. represent a splendid example of this by John Brown of Leverpool. The case is of most exquisite walnut, and the dial and hands and movement are all of high-class workmanship, and yet it was constructed to go thirty hours only.

Returning to London-made clocks, they may be divided into several more or less definite periods, the Ebony and Walnut period, the Marquetry period, the Lacquer period and the Mahogany period. It must be distinctly understood, however, that not only did each period overlap those on either side of it, but it must be recognised that country makers were usually very slow in adopting the prevailing London fashion. This is borne out in the movements, dials and cases.

Another influence which must be considered is the facility for obtaining certain kinds of material and labour in various districts. Mahogany, being an imported wood, was more readily obtainable in towns like London and Liverpool, and other districts relied chiefly on home-grown woods, such as oak and walnut. The first lacquer cases were coated in China, being sent out there for that purpose. London workmen copied them and after a time achieved a certain amount of success, but provincial attempts were naturally poor, owing to the absence of men who had become more or less skilled in the art through prolonged practice.

About 1750 a marked change in fashion occurred and gradually spread through the country districts, namely, the introduction of a dial composed of a single plate of brass, engraved and silvered as shown in Plate XXIV. (*a*). This change was probably not absolutely abrupt. From Plate XXIV. (*b*), which shows an 8 3/4-inch square dial by G. Gibson, of Thetford,* we see that some makers were engraving the corners of their dials. Others were engraving their arches, so it is not to be wondered at that the raised circle was regarded as unnecessary. It had long been known that silver-washed surfaces form an ideal background for engraving; in fact, the Stuart clock shown in Plate VI. was silvered. These dials certainly possess one advantage, viz., the ease with which any one with poor eyesight can see the time. With the introduction of this type of dial came a very much more elaborate minute and second hand, probably with a view to making them match. In Plate XXIV. (*a*), it will be seen that although all three hands are comparatively simple they are of the same design. This type of dial did not persist for many years before the majority of the new clocks had painted iron dials fitted. Cheapness probably influenced makers to adopt the painted dial, but fashion also demanded it. Mr. Prichard has in his possession a fine old chiming bracket clock with brass basket top and dating from about 1698. Until a few years ago this clock had a white painted dial and brass hands. Investigation showed that the old brass dial-base still existed behind the painted one, but the figure ring, corner ornaments, and old hands had been scrapped. Needless to say, the dial has been restored.

(a)

TWELVE-INCH ENGRAVED AND SILVERED DIAL.
JOHN HOLDEN, LINCOLN (?).

PLATE XXIV.

(b)

8 3/4-INCH DIAL, BY GIBSON, THETFORD.

It very soon became apparent that the usual dial feet were unsatisfactory for painted dials, since the paint became chipped round the rivet heads if the clock received a jar in transit. This difficulty was obviated by introducing a stiff iron plate between the dial and movement, which enabled very short feet to be riveted into the former.

With the white dial followed the invariable practice of placing the figures representing the age of the moon on the dial itself instead of sometimes on the dial and sometimes on the moon disc. For the calendar the fan-shaped opening showing a portion of a circular disc, and also a calendar hand and dial as used in the later brass dials, still survived. For a time steel hands remained general, but brass hands were popular for the later movements. Some of the brass hands were extremely well designed and finished with the utmost care, but the majority were comparatively rough stampings produced in shearing dies, the sheared surfaces receiving no subsequent cleaning up or polishing.

Many of the earlier cases used for movements with white dials were of excellent design and of splendid workmanship. So good are many of these old cases that during the last twenty-five years antique furniture dealers all over the country have been buying them and having either old or new brass dials fitted to them. Provided this change is carried out by a man who knows his job the fraud is almost impossible to detect, but in nine cases out of ten the change is plain to be seen by any one who has made a study of old clocks.

By the end of the eighteenth century or the beginning of the nineteenth century artistic clockmaking was nearly extinct in this country. Common looking dials, poor hands and appalling cases became the rule rather than the exception. Probably the worst proportioned cases came from Yorkshire, where there was a demand for something massive. Even the antique and modern furniture dealer, who will buy almost anything at a sale, is reluctant to give more than a few shillings for such atrocities.

Those with an inquiring mind will ask the reason for clocks and cases degenerating, why men departed from the beautiful and adopted the hideous. A possible explanation may be found in the last of a series of articles by Mr. E. Guy Dawber which appeared in the *Architectural Review* of June, 1899. According to Mr. Dawber architects frequently designed furniture for their clients, and he goes on to point out that some of the features of degenerate clocks which he illustrates could only come from a man with an architect's training. On the other hand the writer is informed by Dr. Cranage of Cambridge, an acknowledged authority on art and architecture, that there is not the slightest suggestion of an architect's pencil about the beautiful cases made late in the seventeenth and early in the eighteenth centuries. One very naturally concludes that architects may have been responsible for the degeneracy. If this is so, protect us from architects!

(a)

PETER GUEPIN, LONDON. ABOUT 1700.

(b)

CHARLES GRETTON, LONDON. ABOUT 1695.

* Itis difficult to date this dial. Edward East engraved the corners of his dials in 1665. The engraving in the centre and between the chapters suggest the reign of Queen Anne. The dial evidently had two movements before the writer bought it, without one, thirty years ago. It was apparently made in a small town in Norfolk, so is not particularly old.

THE TYPES OF THE LONG CASES

Any Enthusiastic Admirer of fine paintings knows moments of wrath, albeit silent, when showing some masterpiece to a friend, that friend expresses his lack of appreciation by the comment, 'What a beautiful frame'. And the owner of a fine grandfather clock may experience only slightly less irritation when his 'treasure' is commended with 'Isn't that wonderful wood!' or some similar banality.

Yet both the painting and the clock enthusiast should understand that others, not having their deeper insight and understanding, are affected primarily by what may be termed 'the immediately evident'. This, with a painting, may be a particularly brilliant or unusual tone of colour or finely drawn clouds or it may be the 'beautiful frame'. With a clock it is the intriguing burl (XIX), concentric curves of 'oyster' wood (45), the splendid feather design of mahogany (46), the vivid contrasts of lacquer, the delicate marquetry (XVIII) or any one of the elements of a grandfather which is immediately evident.

Its mechanism may be the acme of fine craftsmanship, the maker may have risen through the years to the Olympus of fame and one of his clocks may be highly valuable in terms of money; but as these factors are normally hidden—and, at best, few have sufficient technical experience to appreciate the mechanism of a clock—approval or otherwise is influenced by what is seen and, to some extent, understood. Therefore, it is natural that the average person should form an opinion of a grandfather clock from the case and the dial.

One late seventeenth-century writer, John Smith, in discussing pendulum clocks refers to the method of 'setting up long swing pendulums after you have taken it from the coffin', the coffin referring to the early type of long wood case, a type which will call forth no feeling of admiration except from one who 'knows' clocks—others, it would 'leave cold' to use an expressive colloquialism. Incidentally, in former times, the word 'coffin' meant a chest or case and is closely related to 'coffer' which formerly had a similar meaning.

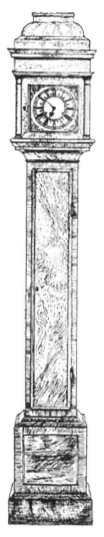

43. *Plain narrow case with plinth below base and small square dial typical of the earliest grandfather clocks*

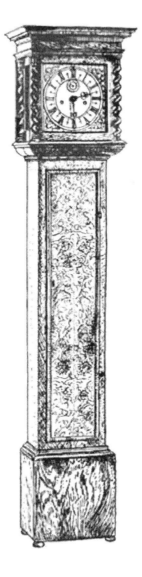

44. Figured walnut case with flat top and spiral columns to the hood. Movement by Thomas Tompion, c. 1695

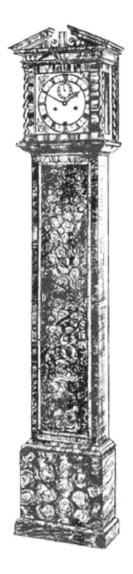

45. Case veneered with oyster wood and cross-banding. The hood has the broken, pediment and columns, c. 1690

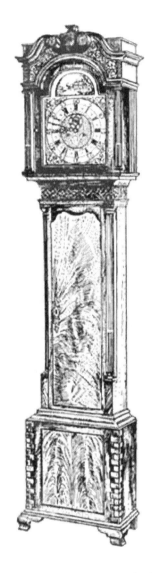

46. *Case veneered with mahogany crotch showing certain features that are often ascribed to Thomas Chippendale, c. 1750*

47. *Irish late seventeenth-century clock in walnut case. The door and base panels are burl outlined with inlay; oval bull's eye showing pendulum bob*

This writer recalls a case of this simple type in the Wetherfield collection dating about 1675. It was slightly over 6 feet high with a small hood, 12 inches square, a long waist only 9 inches wide and a base the same width as the hood. The door which was the full height of the waist was panelled like that of a small cupboard and the sides were also panelled. As time went on all these measurements were increased and there was a noticeable refinement both with the proportions and with other features which added much to the decorative value of these clocks as a piece of furniture.

When clocks were first enclosed in long wood cases, little attention was paid to the actual designs of the cases which were, for some years, of oak. But attractive as this wood can be for furniture, a tall, narrow coffin-like structure such as the early long clock cases, lacks either dignity or beauty. Then, as the grandfather clock came to be regarded as a normal part of the furnishings of a house, there was a demand for a more attractive case and the first move toward replacing the dull woodiness of the plain case was by concealing the humble oak with a veneer of ebony. While these cases gained some popularity in fashionable circles, the application of the ebony veneer was rather a distinction permitted by wealth than any aesthetic improvement.

Some of these ebonized cases have survived to the present time and are of interest as the first indication of the passing of the traditional oak and the coming of the more decorative walnut. The latter may be said to date from the accession of William and Mary in 1688, when oak was superseded by walnut for the cases of important grandfather clocks. And from the former simple panelled style, the cases of the later seventeenth and early part of the eighteenth century developed a sense of elegance which was unequalled at any other time.

If the earlier clocks are not as plentiful as those of the later eighteenth century and if those with movements by certain better known makers command high prices, an appreciable number of examples in cases of the walnut period have come down through the years. Various methods were adopted by the case-makers to achieve decorative surfaces and each of these has its own particular attraction. Where walnut was used, the door and that part of the waist which framed the door as well as the base were veneered with that fine figure known as burl, or burr. This is cut from some part of the tree where distorted fibres have been formed and which gives an indefinable mass of curly and wavy twirls dotted with dark brown spots rather suggestive of tangled wool (44, 47).

As a rule, the base was panelled with burl (sometimes outlined by an inlaid line of lighter wood such as holly) framed by what is known as cross-banding, i.e. the grain of the veneer laid transverse to the general surface; and in these instances the sections of the waist forming the door-frame were usually cross-banded (47). Another strikingly handsome surface was what is known as oyster veneer which is thin slices cut (as a rule transversely) from a sapling or small branch and produces a series of light and dark irregular concentric rings; actually the natural growth rings of the wood (45).

Beautiful as the walnut veneered cases are, they are surpassed for pure magnificence by those decorated with marquetry. And here, without trespassing too far into the fields of technicalities, it may be of interest to give a brief description which, for practical purposes, will define marquetry, parquetry and inlay; for though related, the distinctive differences are not generally familiar. Marquetry is a definite design formed of contrasting woods or other material which are dyed to obtain the various colour tones inlaid in a background of dark veneer and glued to what is known as the carcase which was usually of oak.

Parquetry denotes, more particularly, geometrical pieces of wood fitted together and glued to a bed or core. This is familiar in parquetry floors and the tops of backgammon tables; and it is the general practice to use the term 'inlaid' when the inlaid lighter wood is subsidiary and the greater part of the surface is the darker ground, as for example a mahogany sideboard with drawer fronts and perhaps the top outlined by a narrow strip of some lighter wood, technically known as stringing (47).

Marquetry, which was practised by the Dutch many years before it was known in this country, began to find its way to England in the form of furniture late in the days of Charles II, but it was some years before marquetry furniture was made by English craftsmen. The designs applied to clock cases might be divided into three categories: Geometrical shapes (48, 49) occasionally with a floral panel; floral blossoms and other forms in shaped panels (50, XVIII); and arabesques or other designs, mostly blossoms, which covered the entire front of the case. Marquetry was rarely applied to the sides.

Generally speaking, it may be said that the forms within each of these categories have the same basic origin. The geometric form most commonly used was a circular wheel-like shape with rays or 'spokes' in alternating light and dark wood radiating from the centre; and this may have been an occidental adaptation of the Buddhist wheel which was one of the eight symbols of the promise of happiness. It was variously interpreted in marquetry, but a common form is with wavy 'spokes' in a circle inlaid to suggest a scalloped edge or as a series of tapering 'spokes' similar to what is sometimes called a sunburst and not unlike the small fan-like wheel seen on the towers of wind-pumps. The usual arrangement of the circular ornament with a clock case was twice on the door panel (above and below) and once on the panel of the base, the angles of both panels being 'filled' by a quadrant of the ornament (48). Occasionally, too, a similar but smaller form was added above and below the upper wheel or sunburst.

In other instances, a small oval panel enclosing a vase of flowers was added to the upper part of the door (XVIII); and where this occurs, a wheel-like ornament is often added above and below the panel and a large star-like ornament on the lower part of the door and the base panel. Yet another style of what might be termed the 'detached' marquetry ornamentation as distinct from the 'all-over' style, shows a French influence in the use of scrolls to form a more or less definite outline.

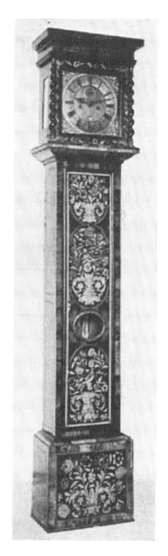

XVIII. Two marquetry cases with the more elaborate floral panels in coloured woods. Both clocks are late seventeenth century, one (left) by Daniel Quare and the other by Henry Massey

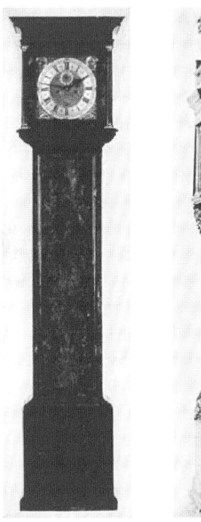

Victoria and Albert Museum

XIX. The case on the left is veneered with burl walnut, the movement by George Etherington, c. 1710; the other is decorated in red and gold on peacock blue lacquer, James Marwick,

c. 1720

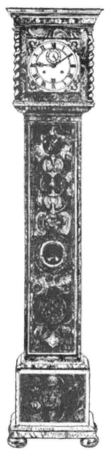

48. *Case veneered with oyster wood inlaid with geometric forms. Circular glass bull's eye in the door. Movement by John Fromanteel, c. 1680*

49. *Case decorated with fan-like designs in marquetry inlaid in oyster wood. Movement by Joseph Knibb, c. 1685 Frederick R. Poke, Esq.*

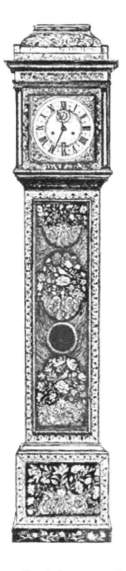

50. Marquetry floral designs in shaped panels

Panels such as have been described are relatively small so that the surface of the walnut in which they are inlaid predominates. With cases where the marquetry design is more elaborate, the panels occupy the whole width of the door and, generally, the panel of the base (XVIII). Cases with what we have termed the 'all-over' ornamentation unquestionably represent the height of the marquetry artist's achievement, as for example, where the design consists of flowers and leaves surrounding a central vase, tiny birds and other motifs in mass. Other designs, even more intricate, are reminiscent of petit point needlework.

Though these clock cases ante-date *The Grammar of Ornament* by more than a century and a half, some of the marquetry designs show a decided resemblance to illustrations in that massive tome. And for an appreciation of the more intricate marquetry work we might well borrow from that volume where, speaking of the renaissance ornamentation, the author says, 'in the majority of which gracefulness of line and a highly artificial, though apparently natural, distribution of the ornament upon its field are the prevailing characteristics'.

There is no question regarding the artistic qualities of the marquetry decoration, but it is undeniably un-English in character. We call for suitable colour as a background to our furniture, but we have a marked preference for quite plain and no great liking for painted or multicoloured furniture, restricting the 'pictorial' additions to paintings, hangings and ornaments.

This native characteristic probably explains the decline of marquetry clock cases which by about 1725 were no longer fashionable. The finely veneered walnut cases remained fashionable, however, and at about this time clock cases were decorated with lacquer (and quasi-lacquer), possibly as a substitute for veneer, an appreciable number of which were made during the reigns of the first two Georges (51, XIX).

We have inserted 'quasi-lacquer' in the foregoing paragraph because the term refers to those often crude forms made of plaster applied on a ground of japan which is nothing more than one or more coats of paint and varnish. True lacquer which originated in China and found its way to Japan is a gum from the Oriental sumach tree, *Rhus vernicifera*, often called the lacquer tree. The gum when exposed to the air dries and becomes so hard that it will resist any solvent and it was this natural product which the Chinese and Japanese and, later, the European workers used for their groundwork.

Incidentally, only this ground-coating is lacquer, and not the pagodas, trees, quaint figures and other ornament which are applied on it.

Lacquered clock cases with movements by English eighteenth-century makers fall into three categories: (*a*) Those made and decorated in Holland and imported to this country; (*b*) those made and decorated by English craftsmen; and (*c*) that much smaller number which were made by English case-makers and sent to the Orient to be lacquered by Chinese or Japanese workers.

Any study of the lacquered cases shows clearly that the earlier examples are superior to those of after about 1740 when the 'novelty' had become widely popular in this country and quality gave place to quantity. In fact, for some years both lacquered and japanned cases were made in larger centres throughout the country and supplied to clockmakers working in smaller provincial towns and villages. And this explains why a grandfather clock movement bearing the name of a maker in an almost isolated rural district is sometimes found in a case more or less (usually less) artistically decorated with lacquer or japan.

More conservative people of the time, however, preferred the 'quieter' walnut which continued to be made for many years after the introduction of mahogany; the same may be said of oak cases which, in country districts, continued to be made until relatively recent times.

51. Repeating bracket clock by John Ellicott, c. 1770. The case is decorated in the Oriental manner on green lacquer

Many early grandfather clocks, mostly of the late seventeenth century, have a small circular (sometimes oval) piece of glass fitted in the door of the trunk. To those who have not journeyed to any extent into the horological realms this is somewhat of a mystery. It is known as a 'bull's-eye' and is more ornamental than useful, for, set at the level of the pendulum bob it was supposed to magnify the bob as it swung to and fro at the back of the glass (47, 48).

Actually the term 'bull's-eye' is associated with an entirely different craft, i.e., glass-blowing; and, as its origin is rather romantic, we might here indulge in a brief digression: Early window glass was known as crown glass which was made by blowing a mass (technically a parison) of molten glass to a large bubble with a blow-pipe. An iron rod, known as a punty or pontil, was then fixed to the bubble on the opposite side to the blow-pipe—the punty serving as a handle for manipulating the bubble, which was cut away from the blow-pipe. By spinning it rapidly like a mop in front of a very hot furnace the bubble burst and after a while expanded to a large flat disc or crown, as it was called. After being annealed it was cut into glass for windows, but that part that had been fastened to the punty was marred by the protuberance or 'bull's-eye'.

Obviously, the part of the disc with the 'bull's-eye' was far less desirable for window glass because, while it allowed light to penetrate, it was not possible to see through it. Consequently, its use was restricted to fan-lights, attic windows and outhouses. They are still to be seen in old houses and a complete 'crown' with its 'bull's-eye' was often used in a circular window frame in the gable ends of outbuildings or other high-up windows.

Before touching on the later eighteenth-century grandfather clock cases, a few semi-historical notes on mahogany wood will not be entirely irrelevant. Mahogany was introduced to this country about 1725–30, but though it became fashionable for furniture almost immediately, it was not used for clock cases until some years later. The reason for this was that until the middle of the century virtually all the mahogany wood was brought from San Domingo, Cuba. This particular wood was hard and dark-coloured with a somewhat monotonous straight grain devoid of any attractive figure, in addition to which it was both scarce and costly. Obviously, figured walnut would be preferred to wood of this relatively dull character, consequently the former continued to be used for clock cases for some time after it had been replaced by mahogany for furniture.

In fact, it was not until other varieties of mahogany were brought from Central America and the West Indies that the figured wood was obtainable and from that time walnut gave place to mahogany for clock cases. Most of the grandfather clocks familiar to us moderns have mahogany cases for the reason that as the eighteenth century advanced and these 'family timepieces' became cheaper, they were available to those with more modest incomes. And if one of the late eighteenth century with an arch dial does not inspire a connoisseur with the same enthusiasm as one with a small square dial in a figured walnut case, it adds much to the attraction of the hall or (across a corner) to the living-room, is an equally good time-keeper and its tick equally mellow and soothing.

To speak of a case as 'walnut' or 'mahogany' does not necessarily imply it is made entirely of either of these woods, but rather that it is veneered with one of them on a carcase, as it is called; the carcase usually being of oak, though many of the later ones made in the country districts are of pine.

As a general rule, in the case of mahogany, the door and base are veneered with the attractive crotch or other similarly decorative figure and the sides with straight grain—crotch is a plume like 'design' obtained from the joint of a limb to the trunk of the tree and is particularly suitable for larger surfaces such as door panels (46). Here again, it is not uncommon to find provincial-made clocks having the entire case veneered with straight grained mahogany and frequently the base will be left plain instead of being panelled by the addition of a moulding.

Cases with certain features such as pillars on each side of the waist or trunk and the hood, scrolled pediments, bracket feet, fretted ornaments (46) and other forms of a

Chinese character are commonly referred to as Chippendale; but while some elaborate drawings of cases are included in Chippendale's book *The Director*, there is no definite evidence that any were made in his shop; nor is it likely that any case-maker of that time would have attempted to produce one of those highly fanciful designs without first considerably modifying the original; and Chippendale's own conceptions would have little appeal in our time.

Thomas Sheraton, the erstwhile preacher who later entered the field of design, also includes some fantastic cases among the drawings in the second edition of his book *The Cabinet Dictionary* which was issued in 1808. No drawings of clock cases, however, appear in the earlier edition of 1803 and, referring to this omission, Sheraton remarks, 'as these pieces are almost obsolete in London, I have given no design of any; but intend to do it in my large work to serve my country friends.' But if his more ornate designs met with no great favour, his influence remains in the mahogany cases delicately inlaid with lighter wood.

Sheraton's comment that grandfather clocks were 'almost obsolete' was close to the truth, because at the time he was writing they had lost much of their former popularity, particularly in the fashionable world of London. They continued to be made in the provinces, and while the movements bearing the names of men in the smaller towns are excellent, there was, at this time, a marked decline in the designs of the cases. This was specially marked in the Midlands where the cases assumed often ungainly proportions and were embellished with a medley of unassociated ornaments.

How ungainly and ugly the cases of the North country clocks eventually became by the early years of the nineteenth century is shown by those usually bearing the 'sign-manual' of a Yorkshire clockmaker. These are unmistakable for they are upward of 8 feet high and 20 inches wide at the waist with the hood and plinth more or less proportionately wide; and this massive structure is often supported on ridiculously small bracket-type feet. These Yorkshire cases are frequently veneered with curly figured mahogany and inlaid, but while they doubtless attained popularity at the time and in the localities in which they were made, that popularity has never extended to the Southern counties, for at no time did these clocks attract much attention in the auction rooms.

We have dealt at some length with grandfather clock cases hoping to afford a more general familiarity with the progressive styles of the different periods. And something relating to the hoods as distinct from the trunk should be added to what has been said.

Before the introduction of the arch above the dial, all dials and consequently the hoods were square, usually with entablature and a flat top (44, 47) but sometimes surmounted by an ornamental pediment supported by spiral (corkscrew) or plain pillars, the spiral type being specially popular during the late seventeenth and early eighteenth centuries (44, XVIII). During the early years of the eighteenth century a dome-like ornament, some of which resemble the inverted bell-top used with bracket clocks, was added above the entablature (50) and this addition was sometimes accompanied by finial ornaments, one at each of the two front corners of the hood. The frieze, i.e. the narrow flat member immediately below the cornice, was quite commonly fretted to allow the sound of the bell to be heard more distinctly, and some of these fretted sections were of brass.

In passing, we might refer to the absence of a door with the hoods of some early grandfather clocks. To-day, when our family time-teller needs winding, we open the hinged door and use the key; but at one time the hood had no door but was made to slide upward in grooves cut in the back of the case, the hood being raised to a convenient height when it engaged with a spring to prevent its sliding down.

Speaking of hoods and winding, here a word of caution: When winding a grandfather see that the lower door is open so that you can watch the weights and avoid straining the gut after the weight has been wound high enough; otherwise it is likely to bring the weight up against the seat board with a jerk. This risk was allowed for by some of the early case makers who, by an ingenious spring gadget, caused the hood when it was down to be locked by the closing of the lower door.

And still another word of advice in the same connection. After winding your grandfather clock do not be satisfied by locking the lower door. Admittedly this protects the pendulum and weights from youthful fingers, but there is a greater risk from adult fingers playing tricks with the hands. So see that the door of the hood is fastened against trespassers. The hood door is not usually fitted with a lock, but one of several simple devices was adopted to prevent its being opened without first unlocking and opening the lower door. This is often merely a square staple or an L-shaped hook on the inside of the hood door, which passes through a slot in the inner frame where it is fastened by a small piece of wood—in exactly the same way as you might put a piece of wood through the staple accompanying a hasp on the shed door in your garden.

.

Printed in Great Britain
by Amazon